Healthy Life In Islam

Based from The Holy Quran and Al-Hadith

English Edition Standar Version

by

Jannah Firdaus Mediapro

2020

Prologue

Healthy Life In Islam Based from The Holy Quran and Al-Hadith English Languange Edition Standar Version.

Islam comes from the root word "sa-la-ma", as do the words Muslim (one who follows the message of Islam) and "salaam" (peace). The root word "Sa - la – ma" denotes peace, security, safety as it does submission and surrender to Almighty God. This security is inherent in the submission to the One God. When a person submits to the will of God he will experience an innate sense of security and peacefulness. He must also understand that God is the Creator of all that exists or will come to exist, and has power over all things. With this surrender and understanding comes peace – real, easily attainable, and everlasting peace.

From the beginning of time, God has revealed Himself through Prophets and Messengers, who have come with one message. Worship God, without partners, without offspring and without intermediaries. The rules and laws were sometimes different, because they were applicable for the people of a particular time or place, but the creed of each Messenger was the same. Worship Me, and your reward will be contentment in this life and in the hereafter. When Prophet Muhammad came, in the 7th century, BCE, his message was slightly different. He called to the worship of the One God, but his call was for all of humankind. The message was now complete and revealed for all places, and in all times.

Islam was completed for the benefit of all who will exist, until the final Day of Judgement. It is not a religion belonging to the Arabs, although Prophet Muhammad, may the mercy and blessings of God be upon him, was an Arab, nor is it a religion for the Asian countries or the third world. Muslims exist in all continents and come from all races and ethnicities. There are Muslims in New York, Sydney, Cape Town and Berlin as well as Cairo, Kuala Lumpur and Dubai. Muslims are as diverse as this magnificent planet. Islam is also not a religion that accepts part time or halfhearted commitment. Islam is a way of life; Islam is a holistic way of life.

When God created the world He did not abandon it to instability and insecurity, quite the contrary, He sent guidance. He sent a rope, firm and steady, and by holding tightly to this rope an insignificant human being can achieve greatness and eternal peace. A Muslim strives to obey God's commandments and does so by following God's guide to life - the Quran, and the authentic teachings and traditions of Prophet Muhammad.

The Quran is a book of guidance and the traditions of Prophet Muhammad explain and in some cases expand on that guidance. Islam, as a complete way of life, stresses the importance of maintaining good health and offers the ways and the means to cope with ill health. The Quran is a book of wisdom. It is a book full of the wonder and glory of God, and a testament to His mercy and justice.

Through His infinite mercy, God has provided us with a holistic approach to life, one that covers all aspects, spiritual, emotional and physical. When God created humankind, He did so for one purpose – to worship Him.

"And I (God) created not the jinn and humankind, except to worship Me (Alone)." (Quran 51:56)

The comprehensiveness of Islam allows every aspect of life, from sleeping and washing, to praying and working, to be an act of worship. One who is truly submitted to God is grateful for the countless blessings in his or her life and wants to thank and praise God for His generosity, kindness and mercy. Prophet Muhammad explained that we should be thankful to God in every situation, whether we perceive it to be good or bad. The reality is that God is just, therefore, whatever situation a believer finds himself in, he knows there is goodness and wisdom embedded in it.

"Indeed amazing are the affairs of a believer! They are all for his benefit. If he is granted ease then he is thankful, and this is good for him. And if he is afflicted with a hardship, he perseveres, and this is good for him." (Muslim)

The life of this world is not stable. Every person goes through stages and phases; happiness is followed by sadness and then relief or joy, ones' faith is strong and unconquerable, and seemingly, for no reason it plummets, next, by the will of God it slowly rises again. Periods of great fitness and health are followed by injury or, sickness, but with each twinge of pain or suffering a true believer feels some of his sins fall away.

"Whenever a Muslim is afflicted by harm from sickness or other matters, God will expiate his sins, like leaves drop from a tree." (Bukhari and Muslim)

Islam teaches us to be concerned, about the whole person. Following the guidance and commandments of God allows us to face illness and injury with patience. Complaining and bemoaning our situation will achieve nothing but more pain and suffering. Our bodies and minds have been given to us as a trust, and we are responsible for them. The guidance of God covers every aspect of life and there are specific ways of dealing with health issues, which we will begin to explore in the next article.

Healing from The Holy Quran

Islam takes a holistic approach to health. Just as religious life is inseparable from secular life, physical, emotional and spiritual health cannot be separated; they are three parts that make a completely healthy person. When one part is injured or unhealthy, the other parts suffer. If a person is physically ill or injured it may be difficult to concentrate on anything but the pain. If a person is emotionally unwell, he or she may not be able to take care of him or herself properly or find their minds distracted from the realities of life.

When speaking to his followers Prophet Muhammad spoke of the strong believer being better than a weak believer, in the eyes of God. The word strong here can mean strength in faith or in character, but it can equally mean health. Our bodies are a trust from God and we are accountable for how we look after our health. Although physical and emotional health is important, spiritual health needs to be the first priority in our lives. If a person is in spiritual difficulty then life can begin to unravel and problems may occur in all areas.

Injury and illness can happen for many reasons, however it is important to acknowledge and accept that nothing happens in this world accept with the permission of God.

And with Him are the keys of the unseen; no one knows them except Him. And He knows what is on the land and in the sea. Not a leaf falls but that He knows it. And no grain is there within the darknesses of the earth and no moist or dry [thing] but that is [written] in a clear record. (Quran 6:59)

This world is but a transient place, beautified for us by the things we covet, spouses, children, wealth and luxury. Yet these are just passing pleasures and temporary joys compared to the contentment and extreme beauty that is Paradise. To help us secure a place in Paradise God places trials and obstacles in our way. He tests our patience and gratitude and provides us with ways and means of overcoming the obstacles. God is also merciful and just, so we can be sure that whatever trials we face God designed them to help us secure a place of eternal bliss. Injury and ill health are trials and tests that we must face with patience, forbearance and above all acceptance.

Accepting a trial does not mean that we do nothing, of course we try to overcome it and learn from it. Accepting means facing the trial patiently armed with the weapons God has provided for us. The greatest of these weapons is the Quran, a book of guidance, filled with mercy and healing. The Quran is not a textbook or book of medicine, but it does contain guidance that promotes good health and healing.

"O mankind! There has come to you a good advice from your Lord (i.e. the Quran), and a healing for that which is in your hearts." (Quran 10:57)

"And We send down from the Quran that which is a healing and a mercy to those who believe..." (Quran 17:82)

There is no doubt that the words and verses of Quran contain a healing for humankind's woes and ills. It was narrated in the traditions of Prophet Muhammad, may the mercy and blessings of God be upon him, that certain verses and chapters by God's will could bring about healing from disease and distress. Slowly over the years, we have begun to rely more on medicines and physical remedies rather then the spiritual remedies prescribed by Islam. If faith is strong and unwavering, the effect of spiritual remedies may be fast and efficient.

From the traditions of Prophet Muhammad comes the story of the man whom the Prophet sent on a mission. He camped close by to some people who did not show him any hospitality. When the leader of the nearby camp was bitten by a snake, they went to Prophet Muhammad's companion for help. He recited the opening chapter of the Quran over the afflicted man and he arose "as if released from a chain".

It is important to seek a cure from the Quran, in the manner prescribed by the Prophet Muhammad, but it is equally important to understand that it is permissible and at times obligatory, to seek help from medical practitioners. Our bodies are ours, only in trust; we are obligated to treat them with respect and to maintain them in the best way. In accordance with the holistic approach Islam takes to health, there is no contradiction in seeking a cure from both medical science and permissible spiritual means.

The Prophet said: "There is no disease that God Almighty has created, except that He also has created its treatment."

He also said: "There is a remedy for every malady, and when the remedy is applied to the disease it is cured with the permission of Almighty God."

Quran is a healing for the body and the soul. Whenever life becomes too difficult or we are beset by injury, illness or unhappiness Quran will light our way and lighten our burdens. It is a source of solace and ease. In the world today many people have untold wealth and luxury but little contentment. Those of us in the West have access to doctors and medicine, to traditional healing, medical breakthroughs and alternative cures but many lives are full of emotional pain and listlessness. What is missing is belief, faith in God.

In the past several decades, it has become widely accepted that religious belief and practices have a significant impact on both physical and emotional health. Medical and scientific research has demonstrated that religious commitment aids in the prevention and treatment of emotional disorders, disease and injury and enhances recovery. Belief in and submission to the will of God is the most essential part of good health care. The words and recitation of Quran can cure hearts and minds, as well as overcome illness and injury, however complete trust in God does not negate the healing effects of medical science provided we use them only in lawful ways. Indeed, God has power over all things, therefore we need to put our trust in Him, develop a lasting relationship with His book of guidance – the Quran, follow the authentic teachings of Prophet Muhammad and seek a cure, wherever it may be.

Diet & Nutrition

Islam is a code of life. Muslims do not practice only during the weekends or festive seasons; rather religion is an ongoing part of daily life. Islam is organised in a spiritual and moral way, taking into account humankind's innate needs and desires. The tenets of Islam are derived from the Quran and the authentic traditions of Prophet Muhammad, known as the Sunnah, These two sources of revelation are a guide, or a manual for life.

Although, it may, at first, seem like a rather strange analogy; let us compare Islam's life instructions with the manual that comes with a computer. Imagine buying a new laptop without ever having seen any of the technological advances of the last several decades. Would you know where the on/off button was? If you managed to turn the computer on would you know how to look after it, do a system restore, run an anti-virus scan, or generally maintain it? Without a manual, the computer would be not much more than a useless piece of technology.

The computer's designers also designed a manual or guide, knowing that without specific instructions the computer would not be put to the best possible use or do what it was designed to do. Technology usually comes with guarantees and warranties that become useless, unless you follow the manufacturer's instructions. Therefore because we want to get the best possible use from our expensive technology we read the manuals and follow the guidelines.

Islam also offers a specific set of instructions that come with a guarantee, a promise of eternal Paradise. There is no 'use by' date on this guarantee and it allows unlimited extensions. If you make a mistake or 'click' the wrong button the instructions clearly advise you how to make amends and restore normality. God designed and created humankind for the specific purpose of worshipping Him and sent Prophets and Messengers with specific guidance to make our task easy. However, without God's guide to life, humankind can become lost and adrift in a world that does not make a lot of sense or offer any real security and contentment. Lives are lived without purpose or meaning and many people eek out an existence that provides little or no real sense of having a *life worth living*.

The traditions of Prophet Muhammad teach us to cherish good health and realise its true value as one of God's countless bounties.

"And when your Lord proclaimed, "If you give thanks, I will give you more; but if you are thankless, lo! My punishment is dire." (Quran 14:7)

Islam's holistic approach to health includes treating our bodies with respect and nourishing them with, not only faith, but also with lawful, nutritious food. A major part of living life according to the Creator's instructions is implementing a suitable diet. Choosing wholesome food and avoiding the unwholesome is essential to good health. God says in the Quran, **"Eat of the good things which We have provided for you." (Quran 2:172) "Eat of what is lawful and wholesome on the earth." (Quran 2:168)**

The Quran contains many verses of advice about healthy eating that relate to the interconnectedness of physical and spiritual health. Encouragement to eat only good and pure food is often combined with warnings to remember God and avoid Satan. Healthy eating not only satisfies hunger but also has an effect on how well we worship.

"O mankind, eat from whatever is on earth [that is] lawful and good and do not follow the footsteps of Satan. Indeed, he is to you a clear enemy." (Quran 2:168)

If one becomes obsessed with food or indulges in too much unwholesome or junk food he or she may become physically weak or distracted from his primary purpose of serving God. On the other hand, if one concentrated exclusively on spiritual endeavours and neglected their health and nutrition, weakness injury or illness would also result in failure to carry out obligatory worship. The guidance found in the Quran and the traditions of Prophet Muhammad advise humankind to maintain a balance between these two extremes.

A healthy diet is balanced with a mixture of all the foods God has provided for His creation. The variety satisfies all the body's needs for carbohydrates, minerals, vitamins, proteins, fats and amino acids.

Numerous verses of Quran mention the foods God has provided for us to nourish and maintain our bodies. It is not an exhaustive list of dietary requirements but rather a general idea of the types of food that maintain a healthy body and prevent illness.

"He created cattle that give you warmth, benefits and food to eat." (Quran 16:5)

"It is He who subdued the seas, from which you eat fresh fish." (Quran 16:14)

"It is He who sends down water from the sky with which He brings up corn, olives, dates and grapes and other fruit." (Quran 16:11)

"In cattle too you have a worthy lesson. We give you to drink of that which is in their bellies, between the undigested food and blood: pure milk, a pleasant beverage for those who drink it." (Quran 16:66)

"There emerges from their bellies a drink, varying in colors, in which there is healing for people. Indeed in that is a sign for a people who give thought. ." (Quran 16:69)

"And it is He Who produces gardens trellised and untrellised, and date palms, and crops of different shape and taste (its fruits and its seeds) and olives, and pomegranates, similar (in kind) and different (in taste). Eat of their fruit when they ripen..." (Quran 6:141)

"...and from it (the earth) we produced grain for their sustenance." (Quran 36:33)

God has also provided us with a list of foods that are forbidden and apart from these everything else is considered lawful.

"Forbidden to you (for food) are: dead animals - cattle-beast not slaughtered, blood, the flesh of swine, and the meat of that which has been slaughtered as a sacrifice for other than God..." (Quran 5:3) "...and intoxicants." (Quran 5:91-92)

While sweets and junk food are not forbidden they must be eaten sparingly as part of a balanced diet, designed to maintain optimum health. Many of the most common chronic illnesses today derive from unhealthy eating habits. Coronary heart disease, hypertension, diabetes, obesity and depression have all been linked to inadequate diets. The traditions of Prophet Muhammad praise moderation as a way of maintaining good health and the Quran stresses the need to strike a balance between any extremes.

True believers need healthy bodies and minds in order to worship God in the correct way. To maintain a sound mind, a pure heart and a healthy body special attention must be paid to health. The heart and the mind are nourished by remembrance of God, and worship performed in a lawful way, and the body is nourished by partaking of the good and lawful food God has provided. Attention to diet and nutrition is a part of the holistic health system inherent in Islam.

Fitness & Exercise

Prophet Muhammad, may the mercy and blessings of God be upon him, said a strong believer was better than a weak believer. He was talking in terms of faith and character but also indicating that physical strength i.e. optimum health and fitness were desirable, providing God gave us the ways and means of attaining such strength. Islam's holistic approach to life and thus health offers us the ability to remain strong and healthy. If God decrees that illness or injury are to be part of our lives then Islam provides us with the ways and means of accepting and even being grateful for the tests and trials that envelope us.

This article, the final in a four part series on Islam's holistic approach to health, will examine what Islam, Prophet Muhammad, and the scholars of Islam have mentioned about fitness and exercise. In a separate series of articles, we will look at how Islam suggests we behave when struck by illness or injury.

Believers in Islam must take care of their spiritual, emotional and physical health. Our bodies, the most complex of machines, are given to us by God as a trust. They should not be abused or neglected but maintained in good order. As previously discussed, diet and nutrition play a big part in maintaining the best possible health, so does a lifestyle incorporating exercise. Islam lays emphasis on a simple diet combined with physical exercise.

Fulfilling the obligations of three of the five pillars of Islam requires that Muslims be of sound health and fitness. The daily performance of five prayers is in itself a form of exercise, its prescribed movements involve all the muscles and joints of the body, and concentration in prayer relieves mental stress. Good health is necessary if one intends to fast the month of Ramadan and the performance of the Hajj (or pilgrimage to Mecca) is an arduous task that requires many days of hard physical effort.

Prophet Muhammad advised his followers, to work, to be energetic, and to start their day early, all of which are conditions for a healthy body. He said "O God, make the early morning hours blessed for my nation." Obesity or an inadequate diet, laziness and weakness are all afflictions for which we will be called to account. Even though preventing illness or injury is often out of our control, there are many conditions brought on or made worse by our own lack of attention to diet and fitness. Prophet Muhammad, may the mercy and blessings of God be upon him, said, "Any action without the remembrance of God is either a diversion or heedlessness excepting four acts: Walking from target to target [during archery practice], training a horse, playing with one's family, and learning to swim."

The Prophet Muhammad and his Companions were naturally physically fit. Life was tougher, long distances were covered on foot, men hunted and farmed their food to survive, and there were no useless recreations to produce laziness and waste many hours of otherwise constructive time. The 21st century contains many distractions and forms of entertainment that encourage laziness and induce ill health.

Although advanced technology has many benefits, it is important that time is not wasted in front of the television screen or game console to the detriment of our health. It has been conclusively proven that obesity in children increases the more hours they watch television. Other studies have indicated that this is equally true for adults. Exercise on the other hand has many benefits.

Exercise increases muscle tone, improves flexibility, enhances endurance, strengthens the heart and fights depression. Exercise also helps achieve significant weight loss. Aerobic exercise fights heart disease and high blood pressure, and reduces the risk of diabetes, while weight training increases muscle strength and reduces fat, increases bone density, fights back pain and arthritis, and improves overall mental health.

Respected Islamic scholar Imam Ibnul-Qayyem stated that movement helped the body get rid of waste food in a very normal way and strengthened the body's immune system. He also stated that each bodily organ has its own sport (or movement) that suited it and that horse riding, archery, wrestling and racing, were sports that benefitted the whole body.

Exercise and fitness play an integral part in the life of a Muslim, however it should not come at the expense of religious obligations, nor should it infringe upon the time spent with family members. In accordance with the holistic approach to life, which is Islam, every thing must be done in moderation. There is no allowance for extreme or fanatical behaviour. Letting an exercise regime or a sport take over your life is against the teachings of Islam that call for a middle path and a balanced approach. Exercise and fitness should also not involve unnecessary mixing of the sexes or wearing clothing that exposes the parts of the body that should be kept hidden.

Islam encourages anything that promotes refreshing the mind or revitalising the body provided it does not lead to or involve sin, cause harm, or hamper or delay religious obligations. The traditions of Prophet Muhammad undoubtedly encourage involvement in sporting activities as a way to promote a healthy lifestyle and encourage brotherly love and family togetherness.

In a narration recorded by Imam Bukhari (a scholar who compiled Prophetic Traditions) it states that "The Prophet passed by some people from the tribe of Aslam while they were competing in archery (in the market). He said to them, **'Shoot children of Ishmael (Prophet) your father was a skilled marksman. Shoot and I am with so and so.'** One of the two teams therein stopped shooting. The Prophet asked, **'why do not you shoot?'** They answered, 'How could we shoot while you are with them (the other team). He then said, **'Shoot and I am with you all."** In another tradition Prophet Muhammad's beloved wife Aisha mentions their love of games and sports. She said, "I raced with the Prophet and I beat him. Later when I had put on some weight, we raced again and he won. Then he said, 'this cancels that (referring to the previous race).'"

A true believer recognises the wonder of the human body and is grateful to the Creator. This gratitude is shown in the care and attention given to maintaining optimum health. Islam's holistic approach to health covers all aspects of the mind, body and soul. A truly health conscious person blends diet, nutrition and exercise with the remembrance of God and an intention to fulfil all their religious obligations.

Why Alcohol Is Forbidden In Islam

Islam's holistic approach to health and well-being means that anything that is harmful or mostly harmful, is forbidden. Therefore, Islam takes an uncompromising stand towards alcohol and forbids its consumption in either small or large quantities. Alcohol is undoubtedly harmful and adversely affects the mind and the body. It clouds the mind, causes disease, wastes money, and destroys individuals, families, and communities. Researchers have proven that there is a strong link between alcohol and gambling. Drinking impairs judgement, lowers inhibition, and encourages the type of risk taking involved in gambling and dangerous activities. God tells us in the Quran that intoxicants and gambling are abominations from Satan and orders us to avoid them. **(Quran 5: 90)**

In Australia, a country with a population of around 20 million, about 3000 people die each year from alcohol abuse while 65,000 others are hospitalised. Studies have consistently revealed a link between heavy drinking and brain damage and around 2500 Australians are treated annually for alcohol related brain damage. Research in the United Kingdom indicates that 6% of cancer deaths are related to alcohol abuse and Harvard Centre for Cancer Prevention says that drinking greatly increases the risk for numerous cancers. Alcohol is considered highly carcinogenic, increasing the risk of mouth, pharynx, larynx, oesophagus, liver, and breast cancers. Drinking alcohol during pregnancy can lead to Fetal Alcohol Syndrome, causing the child to be small at birth, have some facial malformations, small eye openings, webbed or even missing fingers or toes, organ deformities, learning disabilities, mental retardation and much more.

Researchers in Australia have also estimated that 47% of all those who commit violent crimes, and 43% of all victims of these crimes, were drunk prior to the event. Alcohol is responsible for 44% of fire injuries, 34% of falls and drownings, 30% of car accidents, 16% of child abuse instances, and 7% of industrial accidents. Even though it is clear that alcohol is responsible for a great many evils it is legal and even encouraged in most societies. In Muslim countries where alcohol is forbidden many people still find it difficult to resist temptation and fall prey to the disease that is alcoholism. Amazingly even in the light of such startling evidence against alcohol, people around the globe continue to consume alcohol in ever-increasing amounts. Why?

Alcohol is one of the tools Satan uses to distract humankind from the worship of God. God states clearly in the Quran that Satan is an open enemy towards humankind yet by drinking alcohol, we invite Satan into our lives and make it easy for him to distract us from our real purpose in life, to worship God.

"Surely, Satan is an enemy to you, so treat him as an enemy. He only invites his followers that they

may become the dwellers of the blazing Fire." (Quran 35:6)

Alcohol affects the mind and makes sinful behaviour and evil actions fair seeming. It creates enmity and hatred between people, prevents them from remembering God and distracts them from praying, and calls them to participate in unlawful sexual relationships. Alcohol generates shame, regret, and disgrace, and renders the drinker witless. It leads to the disclosure of secrets and exposure of faults.

"Satan wants only to excite enmity and hatred between you with intoxicants (alcoholic drinks) and gambling, and hinder you from the remembrance of God and from the prayer. So, will you not then abstain?" (Quran 5:91)

In pre Islamic Arabia, alcohol use was widespread. To eradicate this evil, God in His mercy revealed the prohibition in stages. First, He made it clear to them that the harm of drinking alcohol is greater than its benefit, next He told the Muslims not to come to prayer while intoxicated and finally, He revealed a verse totally prohibiting alcohol.

"O you who believe! Intoxicants (all kinds of alcoholic drinks), gambling, idolatry, and divining arrows are an abomination of Satan's handiwork. So avoid that so that you may be successful." (Quran 5: 90)

When this was revealed the Muslim citizens of Medina immediately began to destroy and empty their alcohol containers into the streets. Even those who were guiltlessly enjoying cups of wine spat the alcohol from their mouths. It is said that the streets of Medina ran with alcohol. Why then is it so difficult to expunge this evil in the 21st century? Believers today must completely trust God, in the same way that the first Muslims trusted God and understood that He was their only Protector and Provider. All power and strength comes from God and a scourge like alcohol can be eradicated only when those affected by alcohol turn to God with complete submission.

The Quran is a book of guidance sent to all of humankind. It is a set of instructions from the Creator for His creation. If we follow these instructions, our lives will be easy and tranquil, even in the face of disaster and mishap. God links alcohol and gambling to idolatry and declares it filthy and evil; however, He is merciful and generous towards the believers and acknowledges the power of addiction.

Islam is committed to encouraging and facilitating those who wish to repent from evil doing and sinful behaviour. God accepts repentance from those who are truly sorry for their actions and committed to staying away from sin. Muslim communities do not ostracise those who have made mistakes but keep them within the fold of Islam encouraging them to seek the closeness to God that will allow them to leave sinful behaviour. Friends, family, and neighbours do not just look away while a person destroys himself or his family. Islam is a community-oriented faith. There is no place for an individual to do what he wants to do, if it hurts others. Alcohol abuse affects not just the alcoholic but also his or her family, and community. There is great wisdom in the prohibition of alcohol.

Infection Control in Islam

In recent years, health professionals around the world have become increasingly concerned with the spread of infectious diseases. Outbreaks of swine flu, avian (bird) flu, and severe acute respiratory syndrome (SARS) have meant that infectious diseases have taken on a global context and are now on the agenda of world leaders and health policy makers alike. In developed and developing countries, health officials are focusing on infectious disease research and linking it to policymaking and infrastructure.

The scope of infectious diseases is progressively more challenged by globalisation. Easy and frequent air travel allows diseases to spread rapidly between communities and countries. Infectious disease control will continue to be confronted by 21st century issues including global warming, conflict, famine, overpopulation, deforestation, and bioterrorism.

Due to ongoing media attention, most of us are aware of the dangers associated with swine flu and bird flu and in 2003 -2004 the world held its collective breath when 8098 people became sick with SARS, before the global outbreak was contained. These three diseases have led to renewed interest in infectious diseases by the public; however, Gideon Informatics, the world's leading global infectious disease database, has tracked and documented more than 20 major infectious diseases since 1972.

Some basic measures are appropriate when trying to control the spread of any or all infectious diseases.

These include meticulous hand washing, covering the mouth when sneezing or coughing, proper disposal of tissues, staying at home and away from public places, and in extreme cases such as SARS, quarantine. In the series of articles entitled Health in Islam, we explained in some detail that Islam is a religion concerned with creating a community of healthy believers.

Islam is a holistic belief system and it takes into account the physical, emotional, and spiritual well being of individuals and societies. Although care of the individual is important, safeguarding communities, including its weakest members, is of paramount importance. More than 1400 years ago, Prophet Muhammad, may the mercy and blessings of God be upon him, was teaching his followers hygiene practices that are still applicable in the 21st century.

From the traditions of Prophet Muhammad, we find evidence that clearly indicates Islam's stance on coughing and sneezing openly. Prophet Muhammad instructed the believers to cover their faces when sneezing. The most obvious effect of sneezing and coughing without covering the mouth is the spread of airborne bacteria and viruses, in addition, droplets invisible to the naked eye, may fall onto surfaces or other people.

According to the Centre for Disease Control in the USA, the virus that causes SARS is thought to be transmitted most readily by respiratory droplets produced when an infected person coughs or sneezes.

What is known as *droplet spread* can happen when droplets from the cough or sneeze of an infected person are propelled a short distance (up to 3 feet) through the air and deposited on the mucous membranes of the mouth, nose, or eyes of persons who are nearby. The virus also can spread when a person touches a surface or object contaminated with infectious droplets and then touches his or her mouth, nose, or eye. The SARS virus might spread more broadly through the air (*airborne spread*).

Islam is referred to as the religion of cleanliness. **"Truly, God loves those who turn unto Him in repentance and loves those who purify themselves." (Quran 2:222)** In the traditions of Prophet Muhammad cleanliness is mentioned as half of faith, therefore, it is important to keep the body fresh and clean and Islam insists on several practices to facilitate this. The private parts are washed after using the toilet and Muslims must pay particular attention to being clean before praying. They wash their hands, faces, (including rinsing the mouth and nose) arms and feet, a minimum of five times per day. Prophet Muhammad insisted that the believers wash their hands, before praying, before and after eating and upon waking up in the morning.

When trying to stop the spread of any type of influenza, including swine flu and bird flu, the first line of defence is frequent hand washing. Both the World Health Organisation and CDC recommend the following precautions. Cover your nose and mouth with a tissue when you cough or sneeze and dispose of the tissue in the trash after use. Wash your hands often with soap and water, especially after you cough or sneeze. Avoid touching your eyes, nose, or mouth, germs spread that way. Stay home if you get sick. CDC recommends that you stay home from work or school and limit contact with others to keep from infecting them.

Infection control in Islam includes isolation and quarantine. Prophet Muhammad, may the mercy and blessings of God be upon him, instituted strategies that are today implemented by public health authorities. He commanded his followers not to travel to places known to be afflicted with illness and he advised those in the contaminated areas or communities not to leave and spread the disease further afield. He said, **"If you hear that there is a plague in a land, do not enter it; and if it (plague) visits a land while you are therein, do not go out of it"**. He also counseled ill people not to visit healthy people.

During the worldwide outbreak of SARS, quarantine officials arranged for appropriate medical assistance, which sometimes included medical isolation and restricted travel movements. The CDC says isolation is necessary not only for the patient's comfort but also to protect members of the public. Many levels of government around the world are legally able to compel sick, infectious people to remain in quarantine or in isolation in order to stop the spread of disease.

The teaching and principles of Islam are designed to benefit all of humankind. Rules and recommendations for personal hygiene and cleanliness promote the well-being of individuals and communities. Infection control is inherent in Islamic hygiene behaviour. Washing the hands, covering the mouth when sneezing or coughing, voluntary isolation, when one is feeling unwell, and restricted travel is an effective and comprehensive public health strategy. Measures taken in the 21st century to prevent the spread of infections and viruses conform almost exactly to the hygiene and infection control practices taught by Prophet Muhammad.

The Benefits of Honey

The benefits of honey don't just stop at the satisfaction of our taste buds. The powerful healing attributes of honey have long been used to promote health and healing. Both the holy Quran and Hadiths (Prophetic traditions) refer to honey as a healer of disease. In the Quran we read, **"And thy Lord taught the bee to build its cells in hills, on trees and in people's habitations… there issues from within their bodies a drink of varying colors, wherein is healing for humankind. Verily in this is a Sign for those who give thought."**

Additionally, in *Sahih Bukhari* we read that the Prophet, may the mercy and blessings of God be upon him, said: **"Honey is a remedy for every illness and the Quran is a remedy for all illness of the mind, therefore I recommend to you both remedies, the Quran and honey."**

Honey offers incredible antiseptic, antioxidant and immune boosting properties for our body and health. It not only fights infection and helps tissue healing but also helps reduce inflammation and is often used for treating digestive problems such as indigestion, stomach ulcers and gastroenteritis.

Researchers from around the world are discovering new and exciting medical benefits of honey and other healing items produced in the hive such as propolis, royal jelly and bee pollen.

As in all foods, the health benefits of honey largely depend on its quality. Raw honey, the purest form, comes straight from the comb and is placed into the honey extractor or press. This is unheated, unpasteurized and unprocessed honey. Because it is raw, it tends to have fine textured crystals in it that occur naturally when glucose, one of three main sugars in honey, spontaneously precipitates out of the super saturated honey solution. If your raw honey crystallizes, simply place the container in hot water for 15 minutes and this will help return it to its liquid state.

Most honey found on supermarket shelves today is not raw honey, but rather commercial honey which has been heated and filtered so that it has a smoother, more appealing look to it. When honey is heated and processed in such a way it will maintain a long shelf life but the vitamins and minerals which benefit the body's immune system are largely destroyed in the process. As such, it is not as nutritious as raw honey.

Honey varieties differ widely in color, texture and flavor. Comb honey, taken straight from the hive, is the rawest and purest form of honey. Its characteristic hexagon-shaped wax cells that are filled with honey can be chewed like gum.

Liquid honey is the most recognizable and easiest to find. It is pressed from the comb and filtered to remove any particles such as pollen grains, wax or crystals.

Creamed honey, also known as whipped honey, granulated honey or honey fondant, has a smooth and creamy consistency. Cream honey does not drip as does liquid honey, and can be spread easily.

The color of honey, largely determined by the floral source of the nectar from which the bees have collected it, is graded into light, amber and dark categories, with darker varieties being more medicinally potent. In general, lighter honey varieties, such as wild flower honey that has been collected from the nectar of several flowers, have a milder flavor. The darker varieties, such as buckwheat honey, collected from the nectar of the flower of the buckwheat grain, have a more robust and stronger flavor. Approximately 23 common varieties of honey include buckwheat, clover, linden, sage, tupelo and wildflowers.

In addition to carbohydrates, honey contains protein (including enzymes) and amino acids, and is high in vitamins and minerals. Some of the vitamins present in honey are B6, thiamin, niacin, riboflavin, pantothenic acid and certain amino acids. The minerals include calcium, copper, iron, magnesium, manganese, phosphorus, potassium, sodium and zinc. While the amino acid content is minor, the broad spectrum of approximately 18 essential and nonessential amino acids present in honey is unique and varies by floral source. Also present are polyphenols, that can act as antioxidants and play a role in cleansing the body of free radicals and reactive compounds that can contribute to serious illness such as cancer and heart disease. It is believed that honey contains a similar range of antioxidants that are found in green vegetables and fruit including broccoli, spinach, apples, oranges and strawberries.

To fight high cholesterol, look no farther than buckwheat honey. In their 2004 study, Effect of honey consumption on plasma antioxidant status in human subjects, biochemist H. Gross and his colleagues from the University of California, Davis, examined the blood results of 25 participants who were each given four tablespoons of buckwheat honey daily for 29 days in addition to their regular diets. At given intervals, samples taken from them following honey consumption showed that there was a direct link between the subjects' honey consumption and the level of polyphenolic antioxidants in their blood. With its high levels of mineral, vitamin and high antioxidant content, a small amount of buckwheat honey added daily could help lower cholesterol by increasing blood levels of protective antioxidant compounds in the body.

People with diabetes often question whether they can take honey. In his book, The Honey Revolution – Restoring the Health of Future Generations, Ron Fessenden MD says that a tablespoon of honey consists of nearly the same carbohydrate content as a cupful of quartered raw apple, and that a diabetic patient can be assured that consuming honey will produce a significantly lower blood sugar response than an equivalent amount of sugar or other glucose-rich starches. The balance of sugars and the presence of multiple co-factors in honey serve to make this natural food quite different from table sugar, high-fructose corn syrup (HFCS) or other artificial sweeteners.

Dr. Fessenden goes on to say that when consumed regularly over several weeks or months, honey will lower blood sugar and glycated hemogobin levels. He says that generally adding three-to-five tablespoons of honey a day to the diet and eliminating most sugar and HFCS should be recommended to people with Type 2 Diabetes.

For centuries, pure honey has been used in children as a home remedy to help alleviate some of the symptoms associated with the common cold. In the December 2006 study on cough suppressants, Effect of Honey and Dextromethorphan on Nocturnal Cough and Sleep from Penn State College of Medicine researchers compared honey to over-the-counter medicines for symptomatic relief of upper respiratory infection, such as cough, in children 2-to-18-years of age. The conclusion was that honey provided a safe alternative for children. Honey outperformed cough medicine in offering a better night's sleep and reducing severity of cough. Across the board, parents in the study rated honey as significantly better than cough medications or no treatment for symptomatic relief of their children's nighttime cough and sleep difficulties.

The American Academy of Pediatrics and other children's health professionals have raised concerns about common over-the-counter remedies, the safety of these cough suppressant products and whether the benefits justify any potential risks from the use of these products in children, especially in children under 2 years of age. Ian Paul MD, M.Sc., a pediatrician, researcher and associate professor of pediatrics at Penn State College of Medicine and Penn State Children's Hospital said, "Additional studies should certainly be considered, but we hope that medical professionals will consider the positive potential of honey as a treatment..." However, do not feed honey-containing products or use honey for infants under one year of age.

In caring for a wound, the American Journal of Dermatology, in Honey in the Treatment of Wounds and Burns says honey applied topically to a wound can promote healing just as well, or in many cases better than conventional ointments and dressings. Its anti-inflammatory properties reduce swelling and pain while its antibacterial properties prevent infection.

To keep your energy levels up use honey as an excellent source of easily assimilated energy; honey is one of the most effective forms of carbohydrates to ingest just prior to exercise and to replenish your energy levels. The glucose contained in honey is absorbed by the body quickly, giving an immediate energy boost, while the fructose content is absorbed more slowly, providing sustained energy. Honey has also been found to keep levels of blood sugar fairly constant compared to other types of sugar, and it appears to be a carbohydrate source that is relatively mild in its effects upon blood sugar compared to other carbohydrate sources.

Honey has been used for thousands of years to help calm a child's cough, aid in cholesterol relief and digestion and to promote healing of wounds. Honey, in its purest form truly is a head-to-toe cure that will surely be used more and more in the years to come.

Dates Relief of Pain

It is a long-established custom among Muslim parents to put a piece of well-chewed date (or other available sweet fruit) in the mouth of a newborn baby. Muslims do this following the practice of the Prophet Muhammad, may the mercy and blessings of God be upon him, believing him to bee, as the Quran says, sent as a healing and a mercy to mankind. We may infer from the way this custom originated that there is a virtue in it. There is - complimentary to the virtue and pleasure of following the Sunnah (the practice of the Prophet) - placing a `sugary substance' inside the mouth of a new-born baby dramatically reduces pain sensation and heart rate.

An interesting scientific medical study, published in the British Medical Journal (No. 6993, 10 June 1995), proved beyond any doubt the benefit of giving a newborn child sugar, in order to reduce the feeling of any painful procedure like heel pricking for a blood sample or before circumcision.

The study, entitled `The analgesic (pain killing) effect of sucrose in full term infants: a randomized controlled trial', was done by Nora Haouari, Christopher Wood, Gillian Griffiths and Malcolm Levene in the post-natal ward in the Leeds General Infirmary in England.

60 healthy infants of gestational age 37-42 weeks and postnatal age of 1-6 days, were randomized to receive 2ml of one of the four solutions: 12.5% sucrose, 25% sucrose, 50% sucrose, and sterile water (control).

The first group of 30 babies received sugar syrup before a routine blood test (heel pricking, which is usually painful) done to detect jaundice. The other 30 babies were given only sterile water as a control group.

Placing 2ml of a 25% or 50% sucrose solution on the tongue before pricking the heel significantly reduced the crying time, compared to babies who got water. Also, their heart rate returned to normal more quickly. The stronger sugar solution had the greater effect, crying being reduced further with increasing concentration of sucrose. From which we may conclude that sucrose (sugar) placed on the tongue may bee a useful and safe form of analgesia for use with newborn infants.

Blass and Hoffmeyer also showed that 12% solution of inter-oral sucrose significantly reduced the duration of crying in new-born babies subjected to heel pricking or circumcision. This study was reported in The Independent newspaper (Friday 9 June 1995) as well as in the British Medical Journal article.

The practice of the Prophet, upon him b piece, is recorded in the collections of his sayings and reports about him, of which the most revered are the two authentic collections of Al-Bukhari and Muslim:

Abu Buradah reported from Abu Musa, who said:

"I had a new-born baby; I took him to the Prophet Muhammad, who called him Ibrahim. The Prophet

chewed a date then he took it and rubbed the inside of the baby's mouth with it."

There are many other reported incidents like this one.

The date contains a very high percentage of sugar (70-80%); it has both fructose and glucose which have high calorific values, it is easily and quickly digestible, and very helpful to the brain. The date contains 2.2% protein, vitamin A, vitamins B1, B2 ad nicotruic acid (against Pellagra); it has traces of minerals needed for the body such as potassium, sodium, calcium, iron, manganese, copper. Potassium, of which percentage is very high, has been found to be very effective for cases of haemorrhage, such as the occasions of birth or circumcision.

We may note that the Sunnah also commends dates for the breaking of the fast in Ramadan. Dates should be eaten, if available, before the sunset prayer - this is medically and nutritionally the best way and the Sunnah.

The great worth of dates is also indicated in a famous and beautiful passage of the Quran, in chapter named Maryam, verses 25-6:

"And shake towards you the trunk of the palm-tree and it will drop on you fresh ripe dates. So eat and drink and be comforted."

This was the prescription of God, the Creator, for the blessed Virgin Mary at the time of the birth of Jesus, the blessed Prophet of God. It was a prescription to make the delivery easy and comfortable.

As the authors of the medical study referred to intend trying new sugary or sweet substances, we shall recommend that they try dates for the newborn for the relief of pain.

"We shall show them our signs on he furthest horizons and within themselves until it becomes clear to them that it is the truth. Is it not sufficient that your Lord is witness over all things?" (Quran 41:53)

References

Ragab, Ahmed (2012). "Prophetic Traditions and Modern Medicine in the Middle East: Resurrection, Reinterpretation, and Reconstruction".

Rosenthal, Franz; Marmorstein, Jenny (1975). The classical heritage in Islam. Berkeley: University of California Press. p. 182. ISBN 0-520-01997-0.

Prioreschi, Plinio (2001). A History of Medicine: Byzantine and Islamic medicine (1st ed.). Omaha, NE: Horatius Press. p. 394. ISBN 978-1-888456-04-2.

Fazlur Rahman Health and Medicine in the Islamic Tradition: Change and Identity. (New York : Crossroad, 1987)

Rosenthal, Franz; Marmorstein, Jenny (1975). The classical heritage in Islam. Berkeley: University of California Press.

Cyril Elgood (1962) The Medicine Of the Prophet. PubMed Central.

Saad, Bashar; Azaizeh, Hassan; Said, Omar (1 January 2005). "Tradition and Perspectives of Arab Herbal Medicine: A Review". Evidence-Based Complementary and Alternative Medicine.

Lightning Source UK Ltd.
Milton Keynes UK
UKHW051533070521
383207UK00014B/336